Lost CAT

A TRUE STORY
of LOVE, DESPERATION, AND
GPS TECHNOLOGY

CAROLINE PAUL

drawings by WENDY MacNAUGHTON

B L O O M S B U R Y

NEW YORK · LONDON · NEW DELHI · SYDNEY

First published in Great Britain 2013

Text copyright © 2013 by Caroline Paul
Illustrations copyright © 2013 by Wendy MacNaughton

The moral right of the author has been asserted

Bloomsbury Publishing, London, New Delhi, New York and Sydney
50 Bedford Square, London WC1B 3DP

www.bloomsbury.com

A CIP catalogue record for this book is available from the British Library

ISBN 978 1 4088 3557 9

10 9 8 7 6 5 4 3 2 1

Designed by Elizabeth Van Itallie

Printed and bound in China by C&C Offset Printing Co. Ltd

All papers used by Bloomsbury Publishing are natural, recyclable products made from wood grown in well-managed forests. The manufacturing processes conform to the environmental regulations of the country of origin

AUTHOR'S NOTE

This is a true story. We didn't record the precise dialogue and exact order of events at the time, but we have re-created this period of our lives to the best of our mortal ability. Please take into account, however: (1) painkillers, (2) elapsed time, (3) normal confusion for people our age.

SAN FRANCISCO

as seen by

A CAT,

as IMAGINED by A

CAT OWNER

〜〜〜

↖ WATER.
A·K·A
CERTAIN DEATH

1.

One day, I was in a plane crash.

The plane, which I was piloting, was nothing more than sailcloth and aluminum tubing and a lawn-mower engine. It was called an "experimental plane," as if the flying part was just sort of a guess. Which it was, on this day anyway. On this day, it was an experiment that had *failed*.

I crawled from the wreckage dazed and bloody. "Please don't call 911," I said to the first person who arrived. But there was no mistaking the dangling ankle, the misshapen wrist, the blood from my head now soaking my green flight suit, the confusion, and the bits of my experimental plane strewn behind me, like a riot in the last moments of a going-out-of-business sale.

At the hospital, they said, "No internal bleeding or brain damage. Aren't you the lucky girl?" Nurses circled with professional ardor, bearing whirring machines and frowns. Doctors poked and prodded. I was told I had a bad break of the tibia and the fibula.

"The Tibia and the Fibula?!" I said, tasting the blood in my mouth, feeling the bruises on my arm, laughing through my morphine haze. When I explained that those were my cats, the staff just nodded, expressionless; to them, I was just another numbskull hallucinating on a gurney. But it was true. Two thirteen-year-old tabbies, affectionately nicknamed Tibby and Fibby, were now wondering where the heck I was and why I hadn't come home.

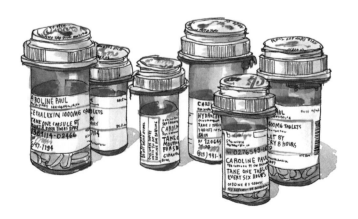

For the next few days, my girlfriend, Wendy, held my hand and assured me that everything was fine. The house was fine, she said. Tibby and Fibby were fine. *You're fine.* She brought me ice and small cups of chocolate pudding, which sat uneaten. She slept in an uncomfortable chair until the nurses told her visiting hours were over, then returned the next day to do it again.

We were in a new relationship, that phase of love that didn't obey any known rules of physics. The past six months had been a stomach-dropping, world-tilting, rainbow-laden, cloud-gilded time, during which we had showed only our perfect selves. That was clearly over. I was overmedicated, wild-haired, unwashed, and fragile, with multiple oozing wounds. There was a bandage for the arm, stitches for the head, and emergency surgery for the left ankle. Wendy ran her hands over my blue paper gown and said I looked beautiful.

The leg was fitted carefully back together. To keep it in place, there was metal scaffolding inside and out. The ankle had been in smithereens, the surgeon told us with an expression that suspiciously resembled glee.

"It looked like those crushed potato chip pieces, the ones you have to tip the bag into your mouth to get to," he said, and mimed the tipping of the bag in case I didn't understand. He shrugged to indicate he couldn't promise anything, despite the joists and girders below my hip. Then he ordered the staff to pump me with morphine. They watched my progress, and finally sent me home to San Francisco.

THE BOTTOM
OF THE BAG.

Tibia and Fibula meowed happily when I arrived. They were undaunted by my ensuing stupor. In fact they were delighted; suddenly I had become a human who didn't shout into a small rectangle of lights and plastic in her hand, peer at a computer, or get up and disappear from the vicinity, only to reappear through the front door hours later. Instead, I was completely available to them at all times. Amazed by their good luck, they took full feline advantage. They asked for ear scratches and chin rubs. They rubbed their whiskers along my face. They purred in response to my slurred, affectionate baby talk. But mostly they just settled in and went to sleep. Fibby snored into my neck. Tibby snored on the rug nearby. Meanwhile I lay awake, circling the deep dark hole of depression.

Without my cats, I would have fallen right in.

Wendy didn't understand the Cat Thing. If Fibby jumped onto her lap, her hands shot skyward in an about-to-be-frisked posture. She patted Tibby on the head as if she was extinguishing a small fire. But she put on her game face when I cooed and babbled like an overwrought aphasic, and she tried to see what all the fuss was about.

But let me introduce you. Here is Fibby, in a typical repose.

FIBBY

See how she rules the world, even while asleep? She was the energetic and sociable one, always eager for human attention. Every lap was designed for her small, round belly, and every nose was a place to put her dainty paw. When my car pulled into the garage, she often leapt from wherever she was and trotted to the entrance, meowing her dismay. Where have you been, her meows seemed to say. And why have you been there so long? Then she would shutter her eyes slowly, wind herself around my legs, and forgive me.

It was hard to believe that Tibby was her brother. He was so anxious and shy. It didn't matter that he was a big cat with large, dark, almond-shaped eyes, so that when he looked at you it was like being stared at by an extraterrestrial. It didn't matter that he had the lope of a tiger and a predator's head, diamond-shaped like a rattler's. In his mind he was a tiny cat, and he slunk around as if the world were going to step on him by mistake. He jumped at loud sounds and ran from strangers. He waited to eat until no one was around. In the backyard he hurried for cover, as if he were on the Serengeti Plain, not a small garden in San Francisco. It had always hurt and puzzled me that my

TIBBY

SAFE ZONE

SLOW MOVING and GENERAL FEAR
OBJECTS

DEFINITE DANGER LIKE WATER and SUCH

DOGS, RACCOONS and LOUD BANGING NOISES

CERTAIN DEATH

love couldn't help him overcome his deep anxieties, that I couldn't reach the vestigial part of him that saw lions and rhinos under every drought-resistant native California plant. But at some point in our years together I had come to accept the simple truth: Tibby was a wimp.

Now, strangely, I understood how Tibby felt. Everything about me was fearful and fragile—not only my ankle but something in my mind, which through

THE DARK

the long summer days now entertained itself with lurid waking dreams, where planes hit the ground with a thud and blood poured onto a flight suit, over and over and over. It could have been the drugs, it could have been post-traumatic stress, or maybe it was simply that in the blink of a left wingtip I had lost that human delusion that the universe was benign and that we were the center of its doting love. In short, I had realized that not everything works out just fine. Things can go to hell fast, and never return to normal.

Weeks went by. Wendy was nursing me heroically, but I was not good company. I trailed a catheter bag and a foul smell. I was filled to the brim with painkillers and regret. I lay supine for hours at a time, watching my leg warily, certain it might do something against my will—perhaps jerk sideways, or head for the floor, or simply break into a million more pieces at the slightest breath of air. I was, in short, getting a little strange. Every day I expected Wendy to lean in, whisper that she'd had enough, and walk out the door. And who would have blamed her? We hadn't been together long enough to justify this kind of burden.

I was confident only of Tibia and Fibula. We'd been together thirteen years, the longest relationships of my adult life. Everything else may have shifted, I thought, as I stared at the ceiling, but the kitties had not, and this was the thing I clung to. Fibby still trotted around the house as if she owned it, and Tibby still lurked in the corners, ready to be petted, but only if Fibby allowed. Tibby and Fibby reminded me that there had been life before angst and injury, and so there would be life after.

But then, a month into my recovery, still bed-bound, depressed, immobile, addled by too much Vicodin, and anesthetized by too much TV, something else happened.

Tibby disappeared.

THE GREAT UNKNOWN

2.

When your cat goes missing, you panic. You imagine catnappers, vivisectionists. You have visions of the hole he is trapped in, the wounds that are keeping him from crawling home.

You cry.

Because I was so helpless, friends rallied quickly. They flyered the neighborhood and knocked on doors. Into every mailbox went the plaintive entreaty LOST CAT, PLEASE CALL, OWNER'S HEART IS BROKEN! Tibby's large, wet extraterrestrial eyes stared from telephone poles and lampposts and trees. Ten days passed. Nothing.

What could have happened? There had been a cat door in my home for thirteen years, through which Tibby and Fibby had come and gone without harm. A narrow street ran down the front of the house, but I had never seen my cats there, and why would they be? Their cat door opened to my backyard and from there the backyards of every house on the block. This long, wide row of fecund foliage offered all a kitty could want—fences and trees to climb, soil in which to roll and snuffle, rodents to catch, grass to eat.

The Indoor-Cats-Only Contingent was now quietly triumphant. They had always scolded that kitties must be kept inside for their own safety. In return,

HI-TECH CAT DOOR

BIG, WIDE, MYSTERIOUS WORLD

I'd scoffed. Sure, everyone would live longer locked in a house, I told them, but we wouldn't be happy or healthy. This was the ongoing debate, each side prancing in their corners, jaws jutted, tones righteous. Now Tibby was gone. If an indoor-only-cat owner had arrived then to shake her bony finger at me, I would still think her misguided. But I nevertheless would have collapsed in tears at her feet.

Desperate, I consulted a psychic. This psychic did not look the way I thought a psychic would. She did not wear large rings or squint into a crystal ball. She sported a stylish haircut and yoga clothes and checked e-mail, which is where I sent her the details of Tibby's disappearance. She responded that she would need a little time to tune in, and so I waited, and soon enough she e-mailed again. Tibby's okay, she wrote, not hurt, and he'll be home by five A.M. on Thursday. This all came through very clearly, she said, and I was not to worry too much about him. In addition, he was being lovingly cared for by nearby children.

From: The Psychic
To: Caroline

Subject: Re:Kitty input
August 5, 2009 6:54:32 PM

Hi Caroline,

- Tibby is safe and not injured. He has been eating too, actually quite well. Some young kids found him, probably over the weekend. Kids between 9-13 years old. They took him home with them.
Yes, I know it may sound strange though I see him in another house with a family (not trapped in a garage). A little girl is now thinking that he is her pet so she's been feeding him and being sweet with him.

We are having a huge full moon eclipse tonight. He may return tomorrow AM when the moon is waning.

Have fun finding him. Bring him home with joy, not fear.

— The Psychic

Ps - this all came to me very vividly. I would be quite surprised if this wasn't Tibby's most current reality.

Children! I thought. He's petrified of children! But I took a deep breath and waited. I admired a psychic who predicted exact dates and times; she seemed so certain. But Thursday came and went. No Tibby. He did not return on the weekend, or the next Monday.

Wendy walked the neighborhood again. She showed photos of Tibby to everyone she saw. People shook their heads with sympathy, said they hadn't seen him, but told her that there was a feral cat colony nearby. Could he be there? I was skeptical. I couldn't imagine Tibby with the rough-and-tumble feline crowd, drinking box wine in corners and throwing gang signs with their paws. Impossible. Wendy wandered the feral cat colony anyway, calling for Tibby, to no avail.

Finally, I put my hands together and asked what God thought. I also asked Allah, Buddha, the Divine Earth Mother, and the Great Vibrant Cosmic Energy. I didn't believe in any of these Things, but I was desperate. "God, Allah, Buddha, Divine Earth Mother, Great Vibrant Cosmic Energy: Where is Tibby? Is he safe?"

There was nothing but silence.

3.

The animal shelter looked like a prison. It had long concrete hallways and heavy doors that rang out when shut. A perky volunteer showed me around. My crutches sounded like hammers thudding on the floor.

The volunteer took me to the cat rooms, which were lined with cages, and stepped back as I peered into each one.

"Tibby?" I whispered. The adult cats were crouched in the back and looked at me without moving. The kittens came forward, but they had drooping tails and mystified eyes. "I'm so sorry," I said to each one. "I wish I could take you home."

I returned to the pound every three days, and every three days it was the same. A volunteer would appear with sympathetic smiles and a bouncy voice.

"I lost my kitty," I would whimper. "He's large, shy,

with wet, extraterrestrial eyes. He disappeared fifteen . . . twenty-one . . . thirty-three days ago."

"Oh, cats," the perky volunteers would respond
knowingly. They would tell me hopeful stories. Everyone had hopeful stories. There were cats who had
been gone for days, weeks, months before returning
home. There were cats who had been found three
thousand miles away, two years later. I listened with
the fervor of the newly evangelized. Clearly the volunteers had some magic that I had lost or never had,
an emotional sturdiness behind their bright smiles.
How else could they stand all this kitty misery?

"You get used to it," one said.

"It's not so bad," said another.

They wore orange smocks and blue paper shoes on
their feet. They cleaned cages and spoke into walkie-
talkies and held sticks with feathers at the end so the
cats could play. I began to love them for their small,
patient smiles, their blue-papered feet, their soft
hearts with tough outer crusts. So I listened raptly
to their tales of kitty intrepidness. Then I went home
and cried.

I e-mailed the psychic again. He's still fine, she

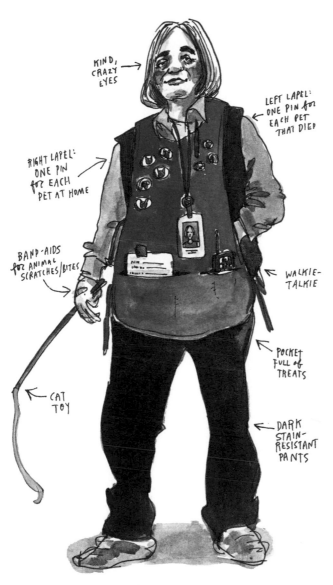

KIND, CRAZY EYES →

LEFT LAPEL: ONE PIN for EACH PET THAT DIED

RIGHT LAPEL: ONE PIN for EACH PET AT HOME →

BAND-AIDS for ANIMAL SCRATCHES/BITES

← WALKIE-TALKIE

← POCKET FULL of TREATS

← CAT TOY

← DARK STAIN-RESISTANT PANTS

ANIMAL SHELTER VOLUNTEER

responded. He'll return with the waning moon. Again I clung to her optimism, the wisdom of her third eye, her good haircut. But the waning moon came and went, and still no Tibby.

And slowly, I knew: A cat like Tibby couldn't survive in the urban jungle. He was too shy, too skittish, with no street smarts, and zero capacity to kick ass. I had to face it; if he hadn't come home, there could be only one reason. Something terrible had happened.

Then, five weeks after he'd disappeared, Tibby returned.

4.

Tibby waltzed into the bedroom late one night. He greeted us with his Pavarotti meow. We sat bolt upright, awakened from sleep. He crawled under a chair.

"Tibby!" I said.

"Tibby!" Wendy said.

Fibby just stared, unsurprised.

"Meow," said Tibby.

I spent the next few days cuddling Tibby and feeling, well, a little indignant. Where had he gone, I wondered, and why had he left? And what was wrong with him now? He was approaching his food bowl with indifference, exhaling a kitty sigh, then walking away.

"He's not eating!" I wailed to Wendy. "He's sick! From being away from home! For so long!"

But when I took him to the vet, he was declared a

half pound heavier. He had a silky coat, said the vet, and a youthful spring in his step.

"That's great," I responded, piqued.

When the relief that my cat was safe began to fade, and the joy of his prone, snoring form—sprawled like an athlete after a celebratory night of boozing—started to wear thin, I was left with darker emotions. Confusion. Jealousy. Betrayal. I thought I'd known my cat of thirteen years. But that cat had been anxious and shy. This cat was a swashbuckling adventurer back from the high seas. What siren call could have lured him away? Was he still going to this gilded place, with its overflowing food bowls and endless treats?

As I spoke (read: ranted), Wendy considered the perfect storm in front of her, of medication, of depression, and of cabin fever, all making landfall on the couch, and nodded with what she hoped registered as sympathy and shared indignation. But the thought bubble that hovered above her head was clear. *What's the Big Deal?* the neon letters shouted. *He's a CAT.*

He was home, she was thinking. Wasn't that good enough?

Well, actually, no.

Wendy abandoned sympathy and tried advice. Perhaps I should lock the cat door for a while so Tibby couldn't wander. I told her I had tried that once, years before. I'd shut him in for a night, and then had lain awake for hours, listening to a loud insistent thudding, which I couldn't identify at first but then realized was Tibby throwing himself against

CAT OWNERS'
THOUGHT BREAKDOWN

the door like a poltergeist. I wasn't going to untrain an old cat, I said. Not now. Besides, I told her, that wasn't the point.

Then for goodness sake tell me, what is the point? screeched the thought bubble, loud enough for my subconscious to hear.

"I can't explain it," I said, my tone haughty, "to someone who hasn't really owned cats."

Where do our pets go and what do they do, when we're not around? And why? Aren't we enough for our furry companions? For animal lovers, these are the ultimate questions. And so began a quest familiar to anyone who has realized that the man in their life is not who he seems: the quest to find out where Tibby had been for those five weeks.

So began Operation Chasing Tibby.

5.

The first and most obvious step in Operation Chasing Tibby was to follow Tibby to his den of iniquity.

"DEN of inIquity," I said with clenched teeth. I could see it: Tibby asleep on a golden pillow, empty tuna cans scattered like rum bottles, young cats lounging nearby. Every so often a palm frond fan of enormous proportions appeared in my mind's eye, floating up and down in considered waves, sending a light breeze across his sun-dappled fur.

"DEN of iNIquity," I said again, re-syllabizing, as if by doing so I was saying something fresh and startling.

"Um," said Wendy. "You mean the place he was for five weeks? But how do you know he's going back?"

I didn't know. Not for sure. But there were signs. He wasn't eating at home, for one. Yet his fur was shiny and his pantherlike girth remained. For two, he had the smug, self-satisfied look of a husband who was getting away with something on the side. I had never experienced that look before, because I'd never had a husband, but I had seen it enough on *One Life to Live* and *As the World Turns* to recognize it immediately.

"Look," I said, pointing. "See?"

Wendy peered at him, but she didn't see.

Then again, she wasn't a veteran cat owner. Of course she would be in the dark.

"Trust me," I told her. "He's enjoying a little hunka-hunka-hunka."

Hunka-hunka-hunka? her face said. But she just nodded, cast surreptitious glances at my med list, and said no more.

Wendy wasn't completely on board with the quest, but she wasn't going to fight it. She had grown fond of Tibby and Fibby. Not fond enough to speak to them in baby talk. Not fond enough to substitute the word "kitty" for the word "cat" in every feline-related sentence. Not fond enough to perseverate over where Tibby might have gone and why. But still, fond. So she wanted to help. But how do you follow a cat? Cats are the slipperiest of domestic animals. Thousands of years of genetic coding has taught them to melt into azaleas, lie motionless behind garden gnomes, glide along fence tops, and slink under benches. Meanwhile, I was on crutches and painkillers.

"We can't go where he goes," she mused. "But technology can."

Which was why I soon found myself at a "spy store," hobbling past shelves of tissue boxes that were really video cameras, past pens that were really tape recorders, past brass knuckles and stun guns and large serrated combat knives. At another time I would have been intrigued by the whiz-bang gadgets. But not today. Today I had a mission.

"I need a tracking device," I said to the young and pimply employee. "You know, something that follows."

"We got that," the employee said lazily, as if a million betrayed wives had been here before me. "You're going to want a Global Positioning System, also known as GPS." He pointed to a cabinet on the far wall and motioned for me to follow.

SPY STORE* NOTES

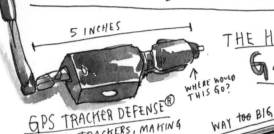

5 INCHES

↑ WHERE WOULD THIS GO?

GPS TRACKER DEFENSE®
CANCELS OUT GPS TRACKERS, MAKING IT IMPOSSIBLE TO FOLLOW. NOTE: KEEP AWAY from CAT.

THE HUNT for
G.P.S.

WAY too BIG

BIG

REAL TIME TRACKER®

USES CELL PHONE TECHNOLOGY, DAY OR NIGHT.

USED for "EXECUTIVE PROTECTION".
PRICE PLAN AVAILABLE.
COMES WITH BELT CLIP.
NOTE: CAT WON'T WEAR BELT. ALREADY TRIED.

2 INCHES

THE 'BIG BROTHER' TRACKER: for MONITORING "NEW TEEN DRIVER" WITH CRASH SENSOR and HELP BUTTON.

3 INCHES

1.5 INCHES

PASSIVE GPS TRACKER®:
PLASTIC, WATER RESISTANT, MOUNTS MAGNETICALLY TO SURFACE, 100 HOURS of TRACKING.
NOTE: CAT NOT MAGNETIC. MAGNETIZE CAT POSSIBLE?

ALSO AVAILABLE:

RECREATES ODOR OF ACTUAL "ROAD KILL", NOW AVAILABLE FOR NON-GOVERNMENTAL SALE. GROSS.

LIQUID ROADKILL

XRAY SPRAY

ENVELOPE XRAY SPRAY: MAKES AN ENVELOPE INVISIBLE SO YOU CAN READ CONTENTS.
SUPER USEFUL IF ANYONE USED MAIL ANY MORE.

and...

B.S. SENSOR

* NOT FOR USE IN THIS STORE.

The glass case we approached was lit like an aquarium. Inside swam GPS devices of every size and shape, bristling with antennae, magnets, screens, and straps. There were GPS units that could be slipped into a spouse's purse, GPS units that could be affixed to the underside of a car, GPS units that could be placed in money bags in the event of an armored car robbery. Informational labels offered long model numbers and promised "one-click satellite overlay" and "integrated antennae" and "flash storage." The young employee lifted a large and heavy-looking box from its shelf and held it toward me with reverence.

"Seventy-two hours of battery life, live tracking through a website, and magnetic mounts," he explained. I looked at the price tag: $1,500.

"I want something a little cheaper," I said and doddered closer to the cabinet. All the units looked much too cumbersome and heavy for a cat to carry. "And it has to be small. Very small."

But each GPS unit he showed me was much too big.

"I can help you better if I know what you need it for," the employee said as he put the last contraption

away. His voice maintained the neutral tone of someone who has been coached about sensitive situations.
But his eyes gave him away. They swiveled up my
crutches, over to the head wound, and back down to
the large contraption on my leg. I knew what he was
thinking. Bad boyfriend? Abusive husband? A con-
frontation with a mistress?

I cleared my throat.

"Um," I said. "Well," I said.

He looked at me expectantly.

"You see," I finally managed, "I need to follow my
cat."

He didn't understand at first, probably because I
was whispering.

"Cat," I said. "C-A-T."

Blank look.

"Consider it a quest to track a very short, very hairy
husband," I said.

Then his eyes lit up. "A cat!" He'd heard a lot of
stories here at the spy store, but he'd never heard this
one. "Wow! Oh, yeah! Well, go on the Internet!" he
cried. "There's so much there. There's definitely going
to be something for a cat, I promise."

Sure enough, my new friend was right. On a strange website full of crude drawings and stiff English, I finally found a very small GPS device. It was made by one man, in his garage, for cats.

Which meant that he was not only a determined engineer; but also a soul mate.

I ordered it.

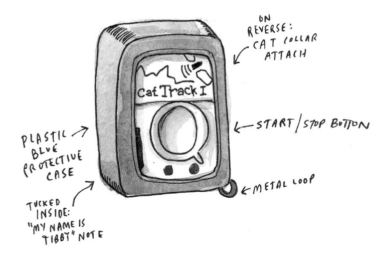

The Cat Tracker arrived. A sturdy, white cube of plastic encased in a blue rubber membrane, it was a little bigger than a Halloween chocolate and about twice as thick, with the same neatness and simplicity. It weighed .75 ounces, at least a third less than any GPS unit at the spy store. There was a button on the front and two lights—one red, one blue—that blinked in various ways, assuring us that what we had was a complicated device that could outwit any medium-size mammalian brain. We went looking for Tibby.

He was sprawled on the rug, snoring. He lifted his head when Wendy and I appeared, not suspicious of our large fake smiles and our slow-motion approach, our murmured nonsense words, the way we looked upward at the ceiling, over at the wall, anywhere but at him. I told him what a pretty kitty, what a smart kitty, what a perfect kitty he was. The unit went on his collar without a hitch.

Tibby was transformed. He was now half cat, half astronaut, with a control panel hanging from his neck, blinking red and blue, lighting up his whiskers. Wendy and I looked at each other, mimed silent

congratulatory speeches, and then peered at Tibby. Would he realize something strange had occurred? But he gazed at us with fondness, unperturbed.

I took some pictures to record the momentous occasion.

He got up and stretched.

He walked toward the door.

He paused at the threshold, then made his way across the hall and sauntered down the stairs.

"Okay," I said. We stood there like parents sending their child off to the first day of kindergarten, proud and forlorn.

"What do we do now?" I said, as his tail disappeared below.

"We wait," Wendy replied.

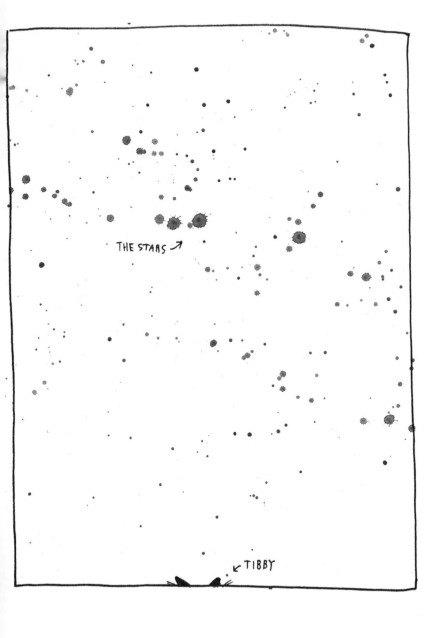
THE STARS ↗

← TIBBY

6.

Twelve hours later, Tibby returned.

"Munya bunya munya munya," I said.* I pulled him onto my lap. "Where the heck have you been, my handsome man?"

As Fibby looked on with disapproval, I gave Tibby a triumphant chin scratch, watched his eyes glaze in pleasure, and unbuckled his collar. The GPS unit slid into my hand. Already I saw the single track that would beam like magic from this chocolate-size circuit board to the computer and then onto its screen—that straight line from our house to the Gilded Place. Once I found the address, I would act quickly, jumping in the car, leaping out at the location, brandishing my crutches like a weapon. I would dangle the GPS unit in

---

*This is baby talk for "Hello, Light of My Life, Kitty of My Soul."

front of the perps and say, "Don't deny it. I have evidence right here."

Fibby waited until I had arranged myself at the computer in the usual way, leg propped to one side, crutches laid against the chair. Then she jumped up and pranced on my thighs like a Lipizzaner. She meowed, as if eager to see the route, though I suspected that she was already in the know, and perhaps had been all along. This theory—that Tibby had told Fibby all about his wanderings—had been greeted by Wendy with a small, unbelieving smile and the rise of an eyebrow.

Crazy cat person, her expression said.

Crazy kitty person, I wanted to correct her.

I tapped at the computer keys, Fibby on my lap. Tibby, seeing that I had been colonized by his twin, walked to his favorite place on the rug. He lay down, oblivious to the fact that he was still the center of my attention.

The screen lit up. I blinked, ready for his track, that one line, straight and true. Instead, here is what I saw:

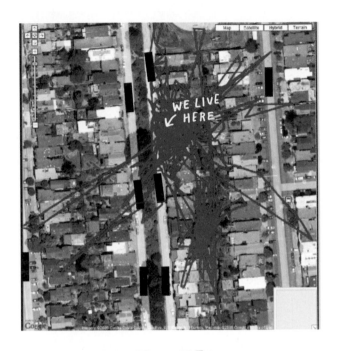

PLEASE NOTE:

■ = STREET NAME REDACTION

THE AUTHOR + ILLUSTRATOR NEED
to MAINTAIN SOME LEVEL of
PRIVACY, NO?

"Holy moly!" I said.

"Holy moly," Wendy said back.

One straight line? No.

The screen looked as if a kindergardener had

been given a Twinkie, and then been let loose with a crayon. It was chaos.

We hung the GPS back on Tibby's collar. He went out, and when he returned hours later, I again slipped the unit from his neck and plugged it into the computer. This time, certainly, there would be a definitive line.

Instead I got this:

I had no idea how to read this riot of feline foot-steps. Should I follow the lines that went westward across the street? Should I concentrate on the nucleus that seemed to stay within our block? And look at that trail that headed east toward the feral cat colony. My head was reeling. I recharged the GPS unit and attached it again to Tibby's collar. With my hands on my hips, and my mouth in a pout, I told him to come back with something a little clearer, for good-ness sake.

This was going to be harder than we'd thought.

7.

"This isn't working," I wailed, pointing to the pile of GPS maps that I had printed and that were now spread on the table before us.

"I think it's working fine," Wendy replied. "You just don't like it that shy, anxious, only-happy-around-you Tibby seems to have such an active social life."

Yes.

"No," I said. "I just think we need to see real evidence. Not just crazy lines on a map."

What we need, I told her, was a camera.

"They don't make cat cameras," Wendy sniffed.

Oh, but they do.

The CatCam, as it was called, arrived in a small padded envelope. It was gray and boxy, and looked as if it had been cobbled together in a basement workshop, which, according to the website, it had. The instructions were in German and English, advising on the Schnellstart of the new device, and after a long period of fumbling, it was ready, programmed to snap a hundred photos, one a minute. That meant that we had under two hours of camera time; enough, surely, to see the perp in flagrante. I imagined a large human visage peering back at me, tuna can in hand. Or perhaps I'd see the furry backside of another cat sharing her food bowl with shy, innocent Tibby.

Sharing that and who knows what else.

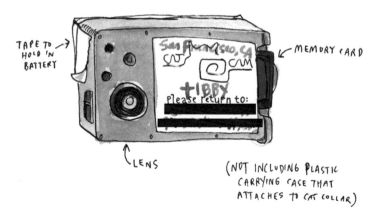

TAPE TO →
HOLD IN
BATTERY

← MEMORY CARD

San Francisco, CA

tIBBY
Please return to:

↑ LENS

(NOT INCLUDING PLASTIC
CARRYING CASE THAT
ATTACHES TO CAT COLLAR)

We found our quarry napping in a sunbeam. He opened one eye when we approached. Again, there was the chorus of excessive praise. There was the circus of innocent expressions. There was the slow-motion approach. Again, the new technological device was attached without a hitch. This time, though, Tibby seemed a little perturbed. What's this? his look was saying. What ridiculous new idea have you latched onto and now clumsily affixed to my neck? We offered another round of sugary exclamations and condolences full of empty promise. Aww, Tibby. Poor Tibby. Good kitty, Tibby.

Tibby huffed and walked down the stairs. As his tail disappeared from view, I said, again, "What do we do now?"

"We wait," Wendy responded.

Wendy walked downstairs a half hour later. Tibby was lying on the sofa. Fibby was nearby, on a chair. The camera was clicking away, that photo every minute. Time was running out.

"Am I ruining the experiment if I put Tibby out-side?" she called up to me. We decided the answer was no.*

Wendy picked up Tibby and brought him to the backyard. "Go to your bimbo!" she told him. Then we forced ourselves to begin our own day. We'd done all we could, we reasoned. Now we had to let the camera do the rest.

My day consisted of lying on the sofa with my leg elevated, imagining where Tibby had gone. Cats have territories, I had been told, and sometimes they are large. True, the GPS didn't back this up. The maps showed that Tibby mostly ran around our block, just as I had always assumed. He ran around like a kitty on methamphetamine, in fact. But every so often, a pink line crossed the street.

I knew that GPS could register "anomalies," espe-cially in an urban environment. Here satellite signals

---

* This is the way we arrived scientifically at this decision:
Wendy: Are you sure we're not ruining the experiment?
Caroline: No, I'm not sure. But that camera is clicking. Can you shoo him, but as if you're not really shooing him?
Some weak sounds of shooing ensue.
Wendy: He's not moving. I'm picking him up.
Caroline: Okay.

become confused, ricocheting off tall buildings, narrow alleyways, abundant foliage, even Tibby's own protruding chin. Could this account for the street crossings? Or was Tibby really wandering two blocks away, maybe more?

I was divided. One part of me wanted Tibby to have remained close by. But this would have meant that he had not only left for five weeks but he had also ignored my frantic calls, and thus me. So, another part of me wanted to believe he had been far away. "Tibby! Tibby!" I had warbled every night until my neighbors hated me. Surely, if he had heard my plaintive cries, he would have returned.

Logic: But he hadn't returned.

Denial: This meant he hadn't heard me. Ergo, he was far away, out of earshot.

Logic: But the GPS doesn't indicate this.

Denial: Wait, he was close by, but he was trapped. Yes, held captive in someone's house, unable to leave.

Logic: Then why did he seem so healthy and happy on his return?

Denial: All right then, so now he's not going back to where he had been. Which was, um, far away.

DID TIBBY HEAR ME?

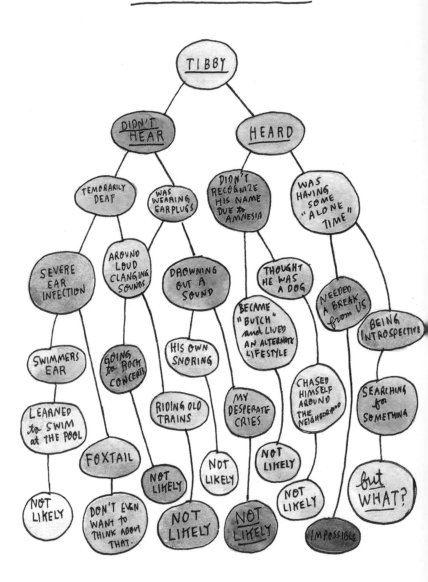

Logic: But he's not eating here. Doesn't that mean he's going somewhere else to eat? Someplace he knows? Someplace the GPS is recording?

Denial: No.

It was becoming clear that I had my own anomalies, pink-lined hopes, hurts, and assumptions. Who could blame me? We'd had thirteen years together! Good years, or so I had thought. Thirteen years of care that included petting, brushing, clean water, space on the sofa, space on the bed, space in my heart, all abruptly forsaken. Thirteen years of food that promised renal health, a shiny coat, strong nails, weight control, urinary tract protection, teeth whitening, nose wetting, ear perking, and tail straightening, now chucked. Thirteen years of love, snubbed.

Why?

Perhaps Tibby had decided to explore the world, like Ernest Shackleton. Wasn't there a kitty version of Shackleton's quest, the planting of a flag in ice, the promise of knighthood, fame, glory, numerous biographies, and a slew of movies? Weren't there continents to cross and mountains to climb, come hell or high water, and finally a kitty hall of fame in which Tibby's large mug would appear, heroic, self-satisfied, triumphant?

Or maybe Tibby had been sowing his kitty oats on a long overdue rumspringa. Like an Amish teenager, he had pitched himself into the wide-open world in search of the life he did not have, only to return home after discovering that heathen sin wasn't as fun as he thought it would be.

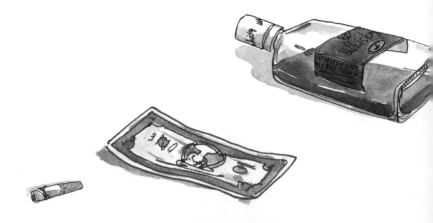

Or had this been a spiritual journey? Tibby was a cat facing the fact that his sprightly years were well behind him, and yet the meaning of life still eluded him. And so, a kitty walkabout.

I was napping, dreaming these half-dreams, when the hero himself returned a few hours later. I shook myself awake and eagerly loosed the camera from his collar.

Here is what I saw:

The perp.

"The perp!" I cried, and Wendy came running. She looked over my shoulder.

"That's me," she said.

The rest of the photos were just as unhelpful. It seemed that once Tibby had been deposited in the yard, he had decided to stay there until the camera ran out of digital space.

There were kitty-under-our-bench shots.

There were kitty-gazing-at-our-sky shots.

WHISKERS

There were kitty-reflected-off-our-glass-door shots.

But no real clues. If he had gone to the South Pole, to the grimy streets of Lancaster, Pennsylvania, or to the Outback, there was no evidence yet.

In every photo, Tibby's whiskers drooped over the frame.

"Those are cute whiskers," Wendy said.

8.

We decided to reprogram the camera's interval time. It would still take one hundred photos, but now they would be spaced five minutes apart. This meant we would be recording eight and a half hours of Tibby's day. Surely we would catch a glimpse of his secret life now.

When I say "we decided to reprogram the camera's interval time," what I mean is "we struggled to understand the directions and hoped we had lengthened the minutes between shots." For something so small and simple-looking, the camera was complicated. And what I mean by "the camera was complicated" is "it kicked our Luddite asses."

I corresponded with the manufacturer. He was German, built these cameras himself by hand, and was eager to help. But the language barrier and my own incompetence meant we fiddled a lot but made little headway. Finally Wendy and I did what most people would at this intersection of hope and desperation. We winged it.

I downloaded the new round of photos with the eagerness of a drug addict who had just scored. But here are a few typical images, caught by the camera:

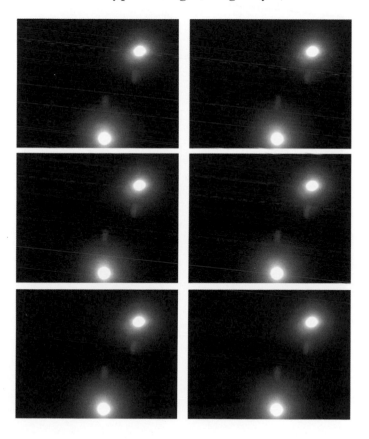

Yes, half of the photos were of ceiling lights. Our ceiling lights.

The other half were no better:

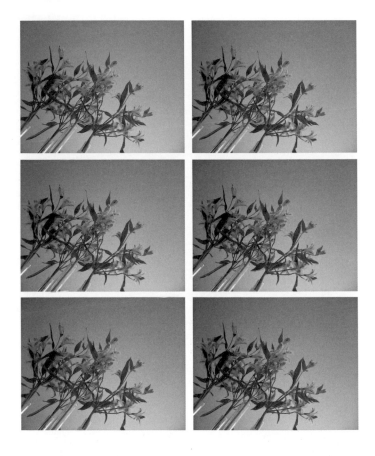

Ensuing rounds were also underwhelming. One showed a continuous montage of a framed picture. This picture hung in direct eyeshot of those prone on the sofa. I could almost hear a kitty snoring sound as I stared at them. My hopes were raised when this was followed by a tilted view of our staircase—Tibby on the move—but were then dashed by a new artistic series: shot after shot of verdant blurs. This was the view from a camera lying on the green down comforter in the guest room.

Wendy squinted at all the photos, nodded, and remarked that at least we had some valuable information.

"What information?" I sulked.

"Tibby sleeps a lot."

Meanwhile, Fibby watched us with baleful eyes. Tibby was getting fussed over, and though it didn't cut into the time logged on her own human-attention-o-meter, she didn't like it.

I loved Fibby but recognized her for who she was: charming, smart, needy, the girl in middle school who ostracized fat kids and boys with bad skin, who whispered loudly behind people's backs, who snickered at the child with the leg braces and the unspecified disease, who spread mean rumors, who instructed others to do so too, who stole lunch money, who lied about what her parents did for a living . . . and who, despite all this, was loved and envied by every classmate.

Tibby was the boy with Coke-bottle glasses whose books were kicked into the mud daily. Fibby was not going to be his friend. Did they even love each other, like twins should? It was hard to know.

Yet.

There were times when I walked into the living room and caught both on the sofa. Upon hearing my approach, their heads would swivel toward me in unison and they would stare with identical expressions, in the way adults gathered at a bridge table with martinis at their elbows might halt their gossip about the floozy down the road at the sudden appearance of a young child—in the doorway, rubbing her eyes, saying she'd had a nightmare, she couldn't sleep, could she have a snack? Amusement, secrecy, conspiratorial smugness—it all seemed to cross over my cats' faces. I would halt in my tracks, overtaken by the sensation that I had not been reading the kitty world around me correctly and that, in some mild way at least, I was being duped. But in the next second, Fibby would rise and stretch and greet me. She would glance at Tibby, a warning look, and Tibby would skedaddle, despite my protests. Then Fibby would climb onto my lap,

WHAT IS SHE TALKING ABOUT?

I HAVE NO IDEA.

curl her plump body into a comma, and gaze at me fondly. She would periscope both ears forward, tilt her head, and finally shutter her eyelids downward, as if they were handkerchiefs falling from a window, thereby completing her dastardly plan. She had now deftly smelted me of doubt, leaving only the pink, bubbly mess that was my heart, and a stupid, glazed, kitty-loving grin on my face.

There were a few more downloads of digital photos, a few more e-mails with the manufacturer, and finally a mea culpa on my part. I removed the camera. It wasn't giving us any answers. Yet the question remained: Where was Tibby going to eat?

Tibby had never been a food-centric animal. But this was San Francisco, and food snobs emerged all the time. Cheap burritos would be abruptly frowned upon; extolled would be meals that were foamed, fluted, flamed, or built, Dalí-like, into strange edifices on your plate. Perhaps this had happened to Tibby. While he was away, he had stumbled upon his inner foodie.

So I whisked the weight-controlling, coat-shining, teeth-whitening, ear-perking, tail-straightening food away. Into a bright, clean bowl I spooned new victuals swimming with plump whole shrimps, fiery orange carrots, and heady aromas. Then I lowered the lights, put on soft music, and crooned that dinner was ready. No, I didn't know who Tibby was visiting. But I was going to beat the bimbo at her own game.

Tibby turned up his nose.

No problem. I opened a different can (creamy duck

ANALYSIS of FOOD CONSUMPTION OVER A 24 HOUR PERIOD

FOOD	CONSUMPTION
A. WATER	MINIMAL. PRIMARILY MEOWED LOUDLY at WATER, SPLASHED WATER WITH PAW, TRACKED WATER OVER HOME
B. DRY FOOD 1. (ORGANIC, $$)	NONE DETECTED
C. DRY FOOD 2. (LOTS OF COLORS CHEAP)	NONE DETECTED
D. FAVORITE WET FOOD	NONE DETECTED
E. TUNA (ALBACORE)	MINIMAL CONSUMPTION
F. CATNIP TREATS	NONE DETECTED
G. FAVORITE TREATS	NONE DETECTED

confit, bright purple eggplant) and crooned once more, but again he turned up his nose. And again. And again.

"No problem," I chirped at each rejection. "A different treat is on the way!"

For days I kept this up, cleaning the bowl and refilling it with something new until my cheeks were numb from smiling, my voice was hoarse from false cheer, and my apron was worn-out. Finally I admitted defeat. Back into the bowl went the weight-controlling, coat-shining, teeth-whitening, ear-perking, tail-straightening food instead.

The GPS returned too. It was all we had until I figured out a new stalking tactic. But the newest map again showed an undecipherable flurry of lines all over the neighborhood. I was stuck: The camera was a wash, providing no clues. And the GPS was incomprehensible.

"Too much information but no way to understand it!" I wailed to Wendy. "We're like Homeland Security!"

"We have made one step forward," Wendy said.

"What!?" I cried.

"You seem less depressed."

It was true. I was getting better. My ankle may have been healing at a glacial pace, but I had a gleam in my eye and a purpose in my heart. True, the gleam was maniacal and the purpose obsessive. But I was slowly, surely, coming back to life.

It was in this state of mind that I decided it was time to do something drastic.

It was time to learn to speak Cat.

"FOOD?"

"I DON'T UNDERSTAND...
FOOD?"

"SERIOUSLY, I HAVE NO
IDEA WHAT YOU'RE SAYING.
FOOD?"

"Oh, MAN. THIS IS GOING
to TAKE A WHILE."

9.

You would think that only a few people would show up to an animal communications class. You would think four or five, ten at the most.

You would be wrong.

In the large college classroom in Marin County, California, fifty people waited eagerly for the lecture to begin.

The room was alive with the sounds of saliva, shifting bodies, and the jingle of metal. That was because thirty attendees had brought their real, live dogs. Mixed in with these sounds was the crinkle of paper; those of us who were dogless clutched flimsy photos of our animals. On my lap lay a two-dimensional Tibby, his large, wet extraterrestrial eyes staring from a partially crumpled head. Next to me, a lumbering black Newfoundland snuffled, looked at me sadly, and then lay down.

I know, I tried to communicate. *Stupid humans.*

Today I had brought two sides of myself to the class: My skeptical side counted the people in the room and added up the money the teacher was making. My earnest side stared at Tibby's photo and told him, "Tonight we're going to have a little chat."

The teacher was a scientist. She said that she approached speaking to animals scientifically. Scientifically, she had come to the conclusion that animal-human communication was well within everyone's grasp.

She said, "Talking to an animal requires only a loving intent, followed by thoughts."

She said, "The thoughts are most powerful if they're in pictures."

She said, "Receiving communication requires an uninhibited mind. Your job is not to filter; it is to recognize and record."

She said, "The first thing that pops into your head is probably from the animal."

She said, "In order to do this, you have to put aside the fact that you think I'm crazy."

Then she asked us if there were any questions. ("Are you crazy?" Skeptical mouthed to Earnest, and laughed meanly.) Yes, there were questions, actually. A young woman raised her hand. Her cat, she said, had passed away recently. How could she speak to him?

"Put all your questions in the past tense," the teacher said, nodding with sympathy.

Someone else asked about talking to coyotes. Another, whether her dog and cat could talk to each other.

The teacher then hovered over a nearby beagle, swooped him off the floor, and cradled him in her arms. He had droopy eyes, a graying muzzle, and a look of resignation. This was her beagle, and we were going to talk with him, she told us, and we all leaned forward so that not a single thought-picture would be missed.

Here's what happened next: The teacher sent the beagle feelings of love. She requested his permission to ask him questions. She asked him to tell us about himself. Or at least that's what was supposed to be happening. From where I sat, she was just staring at him. She may have been secretly looking for fleas.

We were to receive the information the beagle transmitted and write it down. "Free your mind," the teacher reminded us as we waited for word from our canine friend. "Remember," she said, "the first thing that pops in is probably from him."

I wrote: "droopy-eyed, old, carpet, asparagus, brussels sprouts, red jacket."

What did this mean? I had no idea. I did know that we had cooked asparagus and brussels sprouts the night before. I owned a red jacket that I had debated wearing today. We had carpets. Could it be that what was popping into my mind was information not on the life of the beagle but from my own? Yes. So when the teacher asked us to shout out our list, I wisely stayed silent.

Here were some of the shouts:

Loves food!

Loves the outdoors!

Loves dogs!

Doesn't like being held upside down!

The teacher grew visibly excited.

Hangs out on the porch!

Hangs out on grass!

Plays with other dogs!

The teacher clapped her hands happily.

"That was AMAZING," she said. "Everything was spot-on."

"Really?" Skeptical thought to herself. Nearby, a small white dog pranced and tossed her tiny pink nose. The word SPOILED was written in sparkles on her sweater vest. Skeptical tried to raise her hand; she was going to point out that, um, these impressions were generic for all dogs. But Earnest said to stop being a party pooper. Skeptical sulked and kept quiet.

Now it was time to speak to our animal. I stared at Tibby's two-dimensional face in my lap. I set my intention. I opened my mind. I sent feelings of love. I told myself that his answers would arrive quickly and

certainly, and possibly be very odd. I thought, My job is to recognize and record! Finally, I asked, Can we talk?

For a while Tibby just looked back at me, a wrinkle of white along one eye where the paper had been folded.

Suddenly there was a rush of thoughts in my brain.

"Is Wendy staying?" Tibby asked. "Are we getting another comfy chair? Are you going to calm down?"

"Aren't I calm?" I said back in a thought picture.

"Not really. You seem to be worried about things. About the future. The past. What's so great about the future and the past?"

"Well, I don't know," I said. "This is what humans worry about."

"Well, humans are kind of dumb," Tibby responded.

"Hey," I said, adding a thought-picture of OF-FENDED, "this human has fed and cared for you for thirteen years."

Tibby ignored this. Instead I heard, "So, what's going on with this injury anyway. Is it ever going to heal?"

"Heck if I know," I sighed, touched that he cared. Silence.

"Hey, class is wrapping up, so I gotta go," I told him.

"You humans, in such a rush." Tibby said. "You know, this injury might be the best thing for you."

"Really?" I said. But Tibby was already gone.

The teacher told us, "Great job! What a success," but I wasn't so sure. Had I really talked to Tibby, or had I just been talking to myself?

That night, I stared into Tibby's eyes, trying to put all I'd learned that day into play. I stared at him and he stared at me.

Nothing.

Eventually, bored of our game, Tibby put his head on his paws and went to sleep.

10.

Meanwhile, Tibby's twin sister, Fibby, talked to us all the time. Here is a speculative interpretation of her meows:

CAT	ENGLISH
MEOW.	HI.
MEOW.	WAKE UP. WAKE UP. WAKE UP. WAKE UP. WAKE UP. WAKE UP. WAKE UP. WAKE UP. WAKE UP.
MEOW.	I HAVE A GIFT for YOU. SORRY IT BLED ALL OVER THE CARPET.
MEOW.	WELL, THE WATER OUT of YOUR GLASS JUST TASTES BETTER.
MEOW.	WHY ARE YOU EATING ALL THAT CHOCOLATE?
MEOW.	I KNOW THE BATHROOM DOOR IS CLOSED and YOU'RE INSIDE. BUT WHY?
MEOW.	YOU'RE NOT PERFECT but I LOVE YOU ANYWAY.

And yet I missed the one communication I needed to hear:

"I'M SICK."

The day before, Fibby had seemed uncomfortable. But she purred when we petted her and ate what we gave her and batted her kitty eyes. Everything is fine, I said to Wendy. We made plans for the weekend.

It was a beautiful sunny afternoon. We left the house to enjoy it. An excursion! Wendy was right, I was feeling better.

When we returned that evening, we couldn't find Fibby anywhere.

"Fibby?" Wendy called. "Fibby?"

We listened for kitty feet.

Nothing.

I crutched up the stairs. She wasn't on her favorite part of the bed. She wasn't on her favorite part of the rug. She wasn't on her favorite chair. Instead, I found Tibby. He was sitting in the middle of the study, and something in the way he stared at me made my stomach drop.

"Fibby, Fibby," I began to call, at first calmly, but then with rising urgency. There was a puddle of urine on the bathroom rug.

We finally found her in the back of a closet. Her head appeared, then her two front legs. She made it a little way out, then collapsed.

"Oh no," I said, dropping my crutches, getting onto the ground. "No, no, no."

I pulled her onto my stomach. She swayed, couldn't get her balance.

"Fibby!" I cried to her unfocused eyes. "Fibby!"

Wendy stumbled away to call the vet.

We drove to the all-night emergency room. I expected to wait for hours behind the Very Sick, but the assistant peered into the cat carrier, frowned, and whisked it to the back room. The vet came out moments later, with the assistant trailing.

I heard "a large abdominal mass." I heard "very, very sick." They led us to an examining room and spoke in quiet voices, as if we were dangerous.

"I don't understand," I kept saying to Wendy. How could a tumor grow in her stomach without my knowledge? How could I have missed something so big and so bad?

The vet said, "She's bleeding a lot." Then he enumerated the options, none of them good. He spoke slow and low, like a Secret Service agent giving directions on where to place the snipers. Snipers were bad but snipers were necessary, his tone said. He didn't react to the fact that I was weeping. He said, "And all of that might not even work."

"I just don't want her to suffer," I told him through the tissue against my face. "If it was your cat, what would you do?"

Actually what I said was, "Dkpppt jjersss kiii ablutt her sfffffg." But vets are used to translating wracking sobs into a native tongue.

"I'd put her down," he said.

Put her down? I must have misheard. You

put down a foot, you put down a baby, you put down someone you don't like. But all those things can be brought back up. If Fibby was put down, she would be gone forever.

Fibby was slack in her cage. Her pupils were dilated. She whimpered with every breath. We cooed and whispered and ran light fingers along her cheek. Put her down? Only yesterday she had arrived from the pound, it seemed. Only yesterday she had been a ball of fur and ears I could fit into my palm.

Tell me what to do, I tried to communicate. I wanted to do what she wanted to do, and what was best for her. These are two different things in humans, but in animals they're often the same.

Her whimpers continued. I put my head in my hands. I took a deep breath.

"Put her down," I said.

Wendy carried her to a small room. I turned down the lights. Carefully, Wendy put Fibby in my arms. She was so light. How did she get so light?

The vet said, "Tell me when," as if he were pouring coffee. I was weeping, rocking, whispering to the kitty clutched to my chest. I wanted more time, but she was clearly in so much pain.

When? Never, I should have shouted to the ceiling.

Instead, I said, "Now."

Fibby died quickly. Drugs are so efficient.

"She's gone," Wendy said to me. She took Fibby's limp body from my arms.

"Where?" I said, bewildered. "Where did she go?"

Just two days earlier she'd head-butted Wendy's arm, asking for attention. A day ago she'd scrunched up her face like an old man shaving when I'd tickled her chin. Last night she'd eaten her tuna treat.

Wait, we could put the camera and the GPS on her collar.

Wait, please. Please.

But she was gone, and we could not follow.

11.

In the thirteen years we had been together, Tibby had never greeted me upon my return. But that night he was sitting near the top step of the landing. They say that cats don't have many muscles in their face, which is why they seem so much more stoic than, say, a sobbing *Homo sapien*. But the pupils of his extra-terrestrial eyes were dilated. His tail was slack on the floor. His two front paws were together, one slightly in front of the other, like the feet of a ballet dancer about to leap, and the hair on his back was raised. He didn't need facial muscles. It was clear he was asking a question.

Where's Fibby? Where's my twin?

He stared as I leaned on my crutches. He stared as I wiped my cheeks. He stared as I buried my runny nose in my sleeve. When Wendy appeared beside me, Tibby stared at her too. Then he abruptly got up and walked to the den. After a few moments, he came back, glanced at us once, and went into the guest room. There, he peered into corners and under chairs. After covering every crevice, he reappeared. He sat down. He let out one deep Pavarotti meow, so loud and anguished that it startled us both.

"She's gone," Wendy told him.

But he rose to search another room.

For days Tibby looked for Fibby. We know this because the GPS unit remained on his collar. The pink lines now told a story of kitty grief.

First he searched in and around our house, his tracks scribbling circles in the backyard, the living room, and upstairs.

DENIAL*

Increasingly, his path was frantic, furious. He ignored Wendy and he ignored me. He walked with his head down, his tail twitching, his eyes darting about.

---

* Denial is the first stage of Elizabeth Kübler-Ross's five stages of grief. Here the grief-stricken party refuses to believe the terrible news.

ANGER*

Then, map by map, the lines began to fall inward, like a black hole might begin to collapse. He no longer walked the backyard. He narrowed his search to the house itself. Then he simply walked less, lay down more.

---

* In this stage, the party dealing with emotional upset can be angry with himself, and/or with others, especially those close to him.

BARGAINING*

The lines thinned into simple overlapping triangles as his energy drained and it slowly dawned on him: Fibby was nowhere to be found.

---

* The bargaining stage was not as clear-cut for Tibby, though it may have manifested itself when he stayed close to home: "I promise I won't wander anymore if I get my twin back."

Finally, on map 13, all the lines stopped. He wasn't moving now. His track for that ten-hour period was a single, sad trapezoid.

DEPRESSION*

Meanwhile, the humans weren't doing well either. I had a hole in my chest that was Fibby-shaped. I would lie on the sofa and then wake with a start, certain that Fibby was resting her head on my neck.

---

* Tibby had given up.

But there was no Fibby. There was just the phantom imprint of her, her residual weight and heat, a cellular memory.

The cat that left was here. The cat that didn't leave was now gone forever.

Fibby's ashes arrived in a plain wood container. I told Wendy we would pour her into the backyard when I was ready to face it. I didn't tell her it would probably be never.*

Wendy walked around in a daze. She had been indifferent to animals most of her life. But now she was wiping her eyes and talking to herself.

"I miss Fibby," she said to the air.

"I miss Fibby," she whispered to the silence in the house.

"I miss Fibby," she said to me, gripping my arm

---

* When my father moved to San Francisco, his cremated animals came with him, packed in various decorative boxes and baggies. He hadn't gotten around to spreading them, he told me, but he would, very soon. When he died, I gathered all the urns, still unopened, including his. Goodbye Molly, Goodbye Divvy, Goodbye Twiggy, Goodbye Cleo, Goodbye Pru, Goodbye Itty Bitty Kitty, Goodbye Dad. I dipped my hand into the silky remains and scattered them, coughing, onto the flowers, doing for my father what he could not bear to do himself.

and blinking back tears. "It's like a kitty light has gone out."

"Did you mean *cat* light?" I asked.

"No," she snuffled. "I mean a *kitty* light."

12.

Weeks passed. Slowly the maps began to glow again. The pink lines came back to life, larger and wilder, palpitating like EKGs.

ACCEPTANCE*

---

* Tibby was back on the move.

"Now what?" I asked Wendy without much enthusiasm. Would we really continue this quest? Something had gone out of us at Fibby's death.

Yet I wanted a distraction from the lead weight that had been dropped on my heart.

"Maybe a pet detective," Wendy murmured. "Like Jim Carrey."

She couldn't be serious, I thought. A pet detective? They didn't really exist. But I went online and, lo and behold, there were trillions and trillions of pet detectives.

I found America's Most Recognized Pet Detective. I found Electra the Psychic Pet Detective. I found the Law-Enforcement-Based Pet Detective. I found the Top Lost Pet Detective in the United States. There was even a group unfortunately named Missing Pet Detectives, as if the detectives themselves had disappeared.

Eagerly I dug deep into the Web. Here are some things I gathered from my research on pet detectives:

- They like khaki.
- They use FBI profiling techniques.
- They know the most effective poster size and placement.

LAW ENFORCEMENT-
BASED PET DETECTIVE

THE MISSING
PET DETECTIVES

PSYCHIC PET DETECTIVE

THE WORLD'S MOST
RECOGNIZABLE
PET DETECTIVE

- There is an optimal methodology for flyer distribution.
- Most first realized their skill at pet detecting at a tender young age.
- Many use tracking dogs.
- Some also use their sixth sense.
- One or two use ESP.
- They wear floppy hats.

What is an FBI profiling technique? I put this question to a psychic pet detective on the phone who, despite her extrasensory abilities, also used traditional investigation methods. "We hask huestions about the personality of your pet, like police in crime," she replied with an unidentifiable accent, which may or may not have been fake.

"So there is a kitty profile database?" I probed.

"No, no, we use hexperience," she replied. "For ex-hample, I can tell you your cat eez in the past just lost, can't find heez way home. This eez the way of cats."

"He didn't seem lost. He was fat and happy when he returned. He seems to be still going back."

"Yes, yes," she insisted, "Verhy lost."

I sent an e-mail to another pet detective, explaining the situation and requesting his help. Cat! Injury! Cat disappearance! Despondence. Cat return! Cat tracking! Help? The pet detective didn't seem to grasp the predicament. He asked me to fill out an attached questionnaire about my lost cat, and to send money.

My cat was not lost, I re-explained, he was previously lost. But this pet detective was baffled by the meaning of "previously lost," or maybe he simply did not approve of my quest; he had real lost cats to find. We stopped corresponding.

The next detective didn't answer at all. "She thinks you're crazy," Wendy teased.

"I'm not sure an animal private eye has a say about crazy," I retorted.

Finally, I contacted a pet detective who used tracking dogs. She told me that by now Tibby's old scent was gone. Stick with the GPS tracker, she told me, if you really want to do this.

So I hobbled to the library and took out a book called *Secrets of a Private Eye, or, How You Can Be Your Own Private Investigator*. I read it, and then I wrote down what I knew so far:

- Tibby is eating somewhere else.
- This place is the place he had been living before.
- This place is far away, out of earshot. At least two blocks, possibly twenty.

Wendy looked at my list. She was silent for a while. "None of this is proven," she finally said.

"It's called deductive reasoning," I sniffed. I brandished the PI book.

"Well, this, in particular," she said, pointing. "The out-of-earshot, two-possibly-twenty-blocks part. The GPS doesn't indicate Tibby being anything close to twenty blocks away."

"I CALLED HIM EVERY NIGHT," I told her patiently. "If he had HEARD me, he would have come home, clearly. I mean, I was WEEPING. He would never have ignored me, not for five weeks, not when I was WEEPING. This is a THIRTEEN-YEAR RELATIONSHIP. Therefore. Ergo. Clearly." I looked at her as if she had been held back in class a year and I was sorry to be the one who had to tell her.

Wendy seemed momentarily impressed by my tactical reasoning.

"Okay," she finally agreed, in a soft voice used for armed people on ledges. Then she put a hand on my shoulder, looked me soulfully in the eyes, and said, "Perhaps I should take it from here."

13.

It was time for Wendy to step in. Psychics, cameras, an animal communication class, and pet detectives? My ideas had failed us.

"We're going back to the GPS maps," she told me. "We now have twenty-two of them. The answer is in there somewhere."

"But they're just gobbledygook!" I whined.

It was true. Too much information was, well, too much. The volume and geometrics of the pink lines looked more like abstract art pieces than evidence.

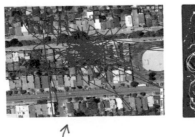

by TIBBY

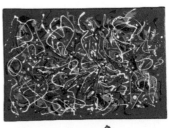

by JACKSON POLLOCK

Wendy patted my knee and agreed that, yes, the lines zigged and zagged like a tachycardic EKG. Yes, GPS was prone to anomalous lines because its signal ricocheted off buildings, foliage, and Tibby's chin. Yes, at this point it was impossible to make heads or tails of the mess of tracks, no matter how hard we stared at them.

"But there's another way," Wendy said.

She ushered me to the computer and settled me in like she was a flight attendant overseeing a minor, ignoring my crossed arms and downturned mouth. She sat beside me. She pecked at the keyboard. The monitor came to life. She summoned the first map, which opened like a firework display, the pink lines, green trees, and gray streets exploding outward, taking over the screen. She called forth the second map. It, too, burst open. Then she began a series of clicks and sweeps with the computer mouse. With the adroit wrist flicks of a ringmaster, she directed lion after lion after lion to jump through a flaming hoop. When she finally sat back, the streets and trees and houses had faded from the second map, but the tracks remained.

She dropped this map onto the first. There were now two sets of tracks superimposed onto our neighborhood.

She repeated this magic with the third and fourth maps, then the fifth and sixth. Clicks and sweeps of the mouse rendered each translucent, stripping it of everything maplike except the pink lines of Tibby's travel. Then each was dragged, one on top of the next, until they were aligned neatly into a cyber-pile.

Suddenly we could see all the tracks on just one tableau.

"Amazing," I cried. If this were a different century, I would have had to turn her into the town council for witchcraft. "But it still looks like a mess."

"Point to the places where there are the thickest lines," Wendy commanded.

I leaned forward, squinted. "There," I said, putting a finger delicately on the screen. "Maybe there"— another finger smudge. "There?"

Wendy nodded, and at each place I had indicated, she dropped a computerized blue dot.

WE LIVE HERE

She then gathered the next six maps, denuded each, and made a new pile. And the next six maps. Then the final four.

Now we had four sets of maps on which we had marked where Tibby went most often. In a final dramatic flourish, Wendy dragged these together and made one master map.

"Holy moly," I said, and put my head in my hands.

"I don't believe it," I whispered in a voice as small as a mouse.

● Outliers   ●●●Concentration over time
(earlier to later = darker to lighter)

14.

Wendy looked at the map in front of us. She looked slowly and carefully, to humor me. She said, "I'm sorry, honey."

"No!" I wailed. "There must be some mistake."

But there was no mistake.

Tibby had not ventured to Antarctica, or to a big city in Pennsylvania, or to the Aboriginal outback.

No.

Instead the map clearly showed Tibby tracks that ended just down the block, again and again and again. Ten houses away.

Ten houses away!

"I don't understand," I whispered with a tubercular wheeze. "This is terrible news."

"This is what you wanted," Wendy said. "To figure out where he had been."

Yes. But also no.*

"And he was so close—isn't that a comfort?" Wendy turned away from the computer and leaned toward me. She tilted her head and widened her eyes to show she was going to be patient and kind, but only for a little while.

"But he heard me calling him each night," I mewled.

---

* Private investigator manual: Do not take on the case if the client seems unwilling or unable to handle information that is humiliating, unsavory, or devastating.

Wendy still didn't understand. I persevered.

"It's as if I phoned you, and you saw my caller ID and you ignored it."

"Well . . ." Wendy looked away. Sometimes she did see my caller ID and ignore it.

"Okay, but it's as if I phoned you, and you saw my caller ID and you ignored it, and so I texted you in capital letters that I was stuck under a train and the left wheel was on my elbow and I needed medical help, and you still didn't pick up the phone."

"It is?" she said. "It's like that?"

Yes, it was. Or maybe it was like a monster had put a large hairy hand into my rib cage and twisted my aorta and pulled out my heart and stomped on it. Maybe that's what it was like.

15.

When the high emotions had subsided, we looked closer at the map. Yes, Tibby had been nearby. But where exactly? There was a three-house-long margin of possibility. Since the backyards abutted one another, there were six houses in all to consider. Wendy labeled this the Suspicious Area.

"We should go door-to-door," Wendy said. "We'll talk to people directly, ask some questions, get some answers."

Wendy had changed in the past few months. She'd gone from being a no-cat person to a cat person. Soon she would be a bona fide kitty person. She had the quest's best interest in mind now. Nevertheless, I vehemently disagreed with her plan. This was a sensitive time. We were getting close; we couldn't blow it.

"We have only one chance to reel them in. After that they get lawyers and clam right up," I said. I had been watching too much TV since my injury. I was now the expert on police procedurals.

"So what's the next step?" Wendy asked.

"We make them nervous," I said darkly. "We use complex psychological pressure."

Wendy leaned in, interested.

"Okay," she said, and waited.

I paused to give her time to get comfortable. Actually, I paused for dramatic effect. Then I said, "We write up a flyer."

Wendy stared at me a long time.

"Do you mean, like the pieces of paper we put in mailboxes and on telephone poles all those weeks he was missing?" she asked.

"Right!" I cried, happy that she was understanding me. "We're going to write a flyer for those six houses and put it in their mailboxes."

"And where exactly does the complex psychological pressure come in?" She was enunciating each word carefully in case I was not just stupid but also deaf.

"Leave that to me," I said, with a wave of my hand. "You'll see."

16.

Here is the note I wrote:

> Dear Neighbor, Thank you for feeding my cat
> Tibby. I'm curious what he's eating. He's fin-
> icky as hell here. Also did you have him for five
> weeks? [Cute picture of Tibby here.]
> Please call me: [phone number here.]
> Caroline

Wendy frowned. After a long pause she said that I sounded angry.

"But I am angry," I said.

I agreed to start over.

> Dear Neighbor, You might remember my cat
> Tibby went missing this summer. I put a flyer in
> your mailbox about it.

Wendy: Take that out. It sounds accusatory and aggressive.

We removed this sentence.

> Luckily he returned after five weeks, safe and sound. However, he is no long eating at my house. I have reason to believe, due to GPS tracking technology, that he has been hanging out in your vicinity and perhaps even eating there.

Wendy: You can't write about the GPS. You sound like a stalker.

Me: I'm not a stalker! I'm a pet owner!

We left this sentence in, but only after an intense battle.

> If you are caring for him personally, I really appreciate it. If he is sneaking your own cat's food, then I apologize. But one way or the other, I'd love to know what food he likes so that I can feed him here at home. You will know him by his blue collar. And the blinking GPS unit that is attached.

Dear Neighbor,

You might remember my cat Tibby went missing this summer,

Luckily, he returned after five weeks, safe and sound.

However, he is no longer eating at my house.

I have reason to believe due to GPS tracking technology that he has been hanging out in your vicinity and perhaps even eating there.

If you are caring for him personally, I really appreciate it. If he is sneaking your own cat's food, then I apologize. But one way or the other I'd love to know what food he likes, so that I can feed him here at home.

Could you please contact me with any information you might have on the matter?

Thank you.

Your neighbor,

Caroline Paul

We eventually removed these last two sentences.

> Could you please contact me with any informa-
> tion you might have on this matter?
> Caroline Paul [phone number]

"It's perfect," I said. "It's firm but friendly."

Wendy was doubtful. But I pointed out that this was the moment to show my hand. It was time to reveal that I had something close to proof on this matter. I wasn't fooling around. I wasn't a weird stalker person. I had GPS.

"Actually, you sound like a weird stalker person," Wendy replied.

"Put it in a nice font," I said.

That afternoon we stuffed a flyer into the mailbox of each house in the Suspicious Area.

"Someone is going to call very soon," I crowed. "You just wait!"

17.

(NOT RINGING)

18.

After five days without a response I admitted that my attempt at subconscious manipulation had failed. Curiously, though, Tibby had begun to eat at home.

"Aha!" I said.

Even Wendy looked a little pleased.

I said, "He's eating here now because the food elsewhere has suddenly been withdrawn. Therefore . . ."

"Therefore we know that the perp *is* in one of those houses!" Wendy finished for me. We grinned at each other.

Finally, some headway.

A few days later, Wendy again argued that it was time to talk to neighbors face-to-face. We would knock, she said, and ask straight-out if the person who answered the door recognized Tibby.

"I hear what you're saying," I said, deftly employing couples-counseling-speak and a generous smile. "And I respect you and all the thought that has gone into that."

Wendy smiled back, not fooled. This was just the sleight of hand that came before the rebuttal.

Rebuttal: "But here's the problem. This is the city. No one answers their door, unless they're expecting a package."

I went on to explain what a ringing doorbell meant in the urban jungle.

> 1. Con artists holding gas cans, with a story about a car, an empty tank, and the ability to pay back whatever you would loan.

> 2. A trio of greasy-haired environmentalists with clipboards and pamphlets, poised to guilt-trip you into donations for whales and trees.

> 3. Imminent home invasion.

In sum, a ringing doorbell signaled someone who had neither your phone number, e-mail, or Twitter

account. Why would you want to speak to them?

Wendy listened with her customary patience, and then she lifted an eyebrow.

"Are you scared to talk to your neighbors?" she asked.

I snorted through my nose to indicate derision and disdain. "Scared! Scared! HA."

Wendy's eyebrow remained skyward.

"No, not scared," I doth protested. "I just have a better plan."

The plan was simple and elegant. Tibby would go door-to-door! After all, Tibby knew the neighborhood. He knew exactly who he was visiting. We would tie a note to Tibby's collar. "Dear Neighbor," it would say. "Are you feeding my cat (this cat)? If so, please call me so I can thank you."

"See, I'm pretending to be interested in food," I said proudly.

"Very smart," murmured Wendy.

We laminated the note with tape and tied it to his collar with a red ribbon. We opened the back door and watched him saunter away.

19.

(STILL NOT RINGING)

20.

If you had asked me about the block I lived on, I would have told you that it was friendly, family-friendly even, but I would have had very little evidence to back this up. The truth was, my neighbors and I simply didn't have a lot of contact. We greeted each other from our sealed cars like parading prom queens, mouthing soundless hellos, contorting our faces into wide smiles, waving enthusiastically as we pulled into and out of our garages. Every now and then one of us might be caught wheeling our garbage can to the curb, and a few hearty words would be exchanged, about the garbage can, about the weather, about nothing in particular. Otherwise, most of my neighbors were strangers with familiar faces.

I'd lived here for twenty years.

GENERAL
TRAVEL PATTERNS
of NEIGHBOR A.

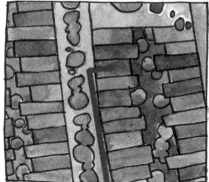

GENERAL
TRAVEL PATTERNS
of NEIGHBOR B.

GENERAL
TRAVEL PATTERNS
of NEIGHBOR C.

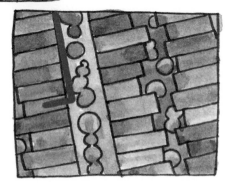

Today I agreed to visit the Suspicious Area. But I had no intention of confronting anyone there. I agreed only to case the joint; all good PIs need to know the lay of the land before they corner the bad guy.

I was full of crime lingo but empty of bravado.

Wendy piled me and my crutches into the car, and we slowly cruised down the street, like cops on patrol. "Here," I said, and we pulled over. For a moment we stared at the three houses the GPS maps had pinpointed.

Wendy said, "It's time to knock on doors."

I'd been backstabbed!

"No!" I cried.

"What are you afraid of?" she asked.

"Afraid!" I scoffed. And then in a smaller voice: "Let's just look."

So we exited the car, to look. Heated whispering ensued.

"Let's just ring the doorbell NOW," Wendy hissed.

"But what if we lose our leverage?" I whined.

"What leverage?"

"The . . . I . . ."

Look, I wanted to say, I'm on crutches. I'm weak,

ungainly, unsure. Why would I want to talk to my neighbors in this debilitated state? I don't even talk to them when I'm healthy. Yes, I'd gone to great lengths to connect to my cat, but when it came to humans in near proximity, I preferred to keep a distance.

It was pathetic that Tibby knew my neighbors better than I did. He knew the state of their fences, the layout of their patio furniture, their gardening skills, the fights they had with their loved ones, the hours they rose, the hours they went to sleep, their tots, their dogs, the aromas of their meals.

We went back and forth like this, me shaking my head and glancing up at the windows, Wendy trying to pump me up but keeping her voice low. She was like a trainer in the corner at a boxing match; in a moment she would be putting Vaseline on my face and stuffing Kleenex up my nose.

"Okay," I finally said.

Wendy rang the doorbell. We peered past the gate, up the garden stairs.

No one is going to answer, I was about to say, when someone answered.

The man seemed to be naked. He was leaning out a side door. I recognized him immediately as the figure in a suit who walked his lanky Doberman past the house in the evenings. Now all we could see was his large white torso, the rest of him disappearing behind the doorjamb. For a moment both Wendy and I were stunned. But his face was not unfriendly.

"Hey!" I said, enthusiastically, sure he was going to

A. draw a gun,

B. yell at us to take our gas cans elsewhere, or

C. pelt us with something he had in his pockets, if indeed he was wearing something, and therefore had pockets.

None of this happened. Instead he looked at us expectantly. "Yes?"

I cleared my throat and held up Tibby's photo. "I'm looking for my cat. I mean, I'm not actually looking for him now. But I want to know if you've seen him."

The man's face softened. He reintroduced himself—we had exchanged names in the past but both of us had forgotten the other's. He'd lived on the block longer than I had, for almost thirty years.

Mr. Naked squinted at the picture and said he had never seen Tibby. He remembered the flyer. He said that since he had dogs, cats didn't usually visit. But the resident two doors up had a cat, he said. Maybe he would know something.

"That was easy," I whispered to Wendy as the door closed behind Mr. Naked.

"That was nice," Wendy agreed. "Now let's try the resident with the cat."

I was sure that this time no one would answer—what were the odds?—but after a few minutes, Cat Resident peeked out from his doorway with the same quizzical look as Mr. Naked had, slightly wary but not unfriendly. And he was wearing clothes. I explained that we were looking for whoever was feeding my cat. I tried not to make it sound like a capital offense, that I was fine that someone was feeding him, in fact, I was happy—no, ecstatic. Over the moon! Wendy knocked me with her elbow and whispered, "He gets it," to redirect me.

Cat Resident said that no, it wasn't him, he didn't

own cats, that it was the guy downstairs, but they're indoor-only cats, and the guy downstairs wouldn't feed someone else's, or at least he didn't think so. Wendy asked if he had any other ideas. Cat Resident, who was now No-Cat Resident, pointed to his backyard and said that Russell, who lived behind him, had cats. "Try there too?"

There was no answer at the lower apartment in the No-Cat Resident's house. "See," I whispered, "people just don't answer their doors in the big city." But Wendy gave me a withering look, so I went quiet. We drove around the block and rang Russell's doorbell, then the next doorbell, and the next, but our luck had run out. No one was answering. But now we had covered part of the Suspicious Area.

"Tomorrow is Saturday," Wendy said. "That's your chance."

"I'm ready," I said.

The day was sunny and warm. I donned clothes that suggested I was a PTA mom on the way to the health food store, took a deep breath, and set out to stalk my neighbors.

I had never noticed how lively my street could be. On this fine day, at least, garages were being cleaned, dogs walked, and kids in strollers rolled to the park as if in some finely staged play. I took another deep breath, mined my list of excuses for one valid enough to quit and return to bed, came up with none, and crutched toward the activity.

The crutches, it turned out, were not a hindrance. Instead, people stopped when I hailed them, gazed with surprise, then advanced, seemingly ready to open any door or carry any grocery bag or call an ambulance. To each I offered Tibby's photo and told my convoluted sob story as quickly as possible. Cat lost! Cat returned! GPS! I ended with this: Seems he was in this area. Have you ever seen him?

At first I didn't have any luck. No one recognized Tibby. But I did have lots of lovely conversations with sympathetic people who crumpled into their How Cute! face at his photo and listened to the story of my quest with interest.

Hello Alastair! Hello Daphne! Hello John! Hello Lorraine! Tibby had now introduced us. I had finally spoken to my human neighbors. Yet Tibby had never been seen, though the GPS showed that he'd been crisscrossing these yards every day for months, probably years. Humans, it was becoming clear, were oblivious to much of their surroundings.

Finally, my neighbor Alastair had a suggestion. "You might try here," he said, and pointed to a house I had seen often; it was smack-dab in the middle of the GPS activity.

"Really?" I said.

"Yes. They once had fifteen feral cats in their house. They're real cat lovers."

Or cat stealers, I thought.

There was no doorbell, and no way to knock on the iron gate, so I stood outside on the sidewalk for a few minutes, gathering the remnants of my courage. I called up to the house.

"Yoo-hoo! Hello?"

There was no answer or movement. I waited anyway. I had had a nice day in my neighborhood, and calling "yoo-hoo" was starting to suit me. I didn't have to wait long. A roly-poly man with a Santa Claus beard, wearing pajama bottoms and a T-shirt, appeared above me on the threshold of his front door. He walked down the outside stairs and squinted at me through the gate.

"Yes?"

I held up Tibby's photo.

"I know that cat," he said.

At home, I was in a tizzy. "I found the perps!" I told Wendy. "Their names are —— and ——!* They live in the Suspicious Area! They once had fifteen feral cats! They were feeding Tibby, they told me!

---

* The names have been redacted because everyone is innocent until proven guilty. From here on in, they will be known as the Cat Stealers.

And that's code for 'we locked Tibby in our closet!' It's lucky he escaped!"

Wendy, calm-headed and reasonable, waited for my rant to end. Then she said, "Let's have them over for tea and find out what really happened."

21.

When you have cat stealers over for tea, you clean the house, buy bagels and cream cheese, and try to figure out how to trap your guests in a lie. You make a list of questions that will ferret out the truth like a drug test ferrets out certain chemicals.

> 1. When you picked up Tibby and brought him into your house and locked him in there for five weeks, did you wonder why he had a collar on him with a name, an address, and a phone number?

2. When you heard a cracking, warbling female voice calling out the name "Tibby" each night, which matched the name on said collar, did you consider there might be a connection?

3. When Tibby seemed sad and withdrawn and homesick, clearly missing his owner, did you
    a. brainwash him
    b. feed him crap food to make him forget his sorrows
    c. ignore him and the facts
    d. all of the above

Wendy said I couldn't use any of these questions.

"Be nice," she said. "We don't know anything yet. We've let the GPS unit, the camera, the animal communicator, the notes, the flyers, and the pet detectives talk. Isn't it time to let our neighbors have their say?"

So I jotted down a few things that we wanted to cover:

    Children in household (per the psychic's reading)?

Address and phone number on Tibby's collar—
     readable?
THREE FLYERS IN MAILBOX
Various objects hanging from Tibby's collar—
     comments?

Wendy looked at this list and agreed to cover these
points. She told me not to interfere; she'd handle the
interview from here.

The Cat Stealers arrived for tea. I peered at them
with narrowed eyes. This is what they looked like:

CAT STEALER A

CAT STEALER B

Yes, they looked like nice people. They even proffered a gift: a leafy sprig of catnip, which they said they grew in their yard.

Wendy was pleased, but I was not fooled. Wasn't growing catnip in one's yard the kitty equivalent of giving candy to children? Yes, it was.

The Cat Stealers were happy to see Tibby, who was lying on the sofa and raised his head when they approached. Aha! This was the true test. Tibby's kitty brain was whirring, trying to place where he had seen these two. Once everything clicked, he would flee from his former captors.

He let himself be petted.

Cat Stealer A must have seen my shocked face. "Cats like me," he said. "I'll go someplace and be sitting, and if there's a cat in the neighborhood, it'll find me."

"So Tibby found you," Wendy said softly. She glanced at me to make sure I was hearing this. They were not Cat Stealers, her look said. They were Cat Whisperers.

Here is their story:

Tibby had shown up one day last summer. He had waited in the shrubbery until the other strays ate the food the Cat Stealers routinely put out; only then did he approach. No, they had never brought him into their house, or even petted him. He always fled when they got too close. If there was an address on his collar they couldn't get close enough to read it. Later, they noticed a large blue box under his chin. They did not have any children. They didn't realize Tibby was missing or lost. They thought he was just there to snack.

But what about the three flyers we'd put in their mailbox?

They had never seen them.

None of them?

None.

The Cat Stealers seemed anxious to help. They discussed it back and forth and finally supposed that the downstairs tenant had disposed of the flyers before bringing up the mail.

Really? Really? Could it be this simple—a skittish kitty, an unreadable collar, a careless tenant?

I wasn't supposed to ask any questions, especially accusing ones. But I couldn't hold back.

"Well, if he wasn't in your house, where did he sleep?" I sniffed. "He certainly wasn't at home."

"My guess is that he was taking a meal from us and then going to the run-down *banya* next door," Cat Stealer B said, and explained that the neighborhood had once been filled with Russian immigrants. It was not uncommon to find the detritus of their *banyas*, Russian saunas, listing into the weeds.

I was hardly listening. Instead I was making a mental checklist. On one side of the paper that floated in my brain was the word "Crazy." On the other was "Not Crazy." The list under "Crazy" was long.

> Feeds strays ☑
> Grows kitty drugs in backyard ☑
> Rents to unreliable tenant ☑

"He must have slept in the banya," agreed Cat Stealer A. "Lots of cats go there."

Tibby had been living in a San Francisco bathhouse? This was too much.

THE
BANYA

"Didn't you have fifteen cats once?" I cried, and then shot Wendy a triumphant look.

The look said, Crazy cat people.

It said, Fifteen cats!

It said, We can't trust them.

"Oh yes," Cat Stealer B said, unfazed. "We had a neighbor who had a whole colony of cats in her backyard. She was very old, only spoke Russian, partly blind, completely deaf, very wary of people. We tried to convince her to bring us the kittens. Then we would spend three or four weeks with each one, play with it, love it, before adopting it out. I would sometimes cry when one was adopted. And we'd tell everyone, 'If you don't want the cat anymore, bring it back to us. Don't abandon it at the pound.'"

Fifteen cats ☑

Hangs out with blind hermit ☑

"We used to keep files on each one. We once adopted out twenty-five in a year," added Cat Stealer A.

Files on each cat ☑

Blackie, Rippy, Nessie, Kbar, Elliott, Buford, Princess, Chloe, Jones—these were just some of the cats our neighbors had either owned or watched over. There had been an especially beloved cat named Lord Brandoch Daha, after a character in a novel. "He used to wander, so I would go to the backyard and call him," Cat Stealer A said, pausing at the memory. "Some people thought I was some sort of weird Christian. There I was yelling, 'Lord! Lord!' at the sky."

They no longer had so many cats in the house. Now there was only one, a big, gentle Maine coon. But they kept food in their backyard, they explained, because there were so many strays and abused animals.

I ignored the pang I felt at the mention of strays and abused animals.

"What kind of food?" I asked, thinking, *Food that was drugged?*

"Friskies," they said.

"Oh, Friskies," I said and raised my eyebrows at Wendy, to be translated as "Friskies is Halloween candy for cats."

Friskies ☑

The list under "Not Crazy" was suspiciously empty. If you looked hard enough, there was some small type. It read:

Adopts out kittens ☑

Gives freely of food to strays and abused animals ☑

Takes care of hermits ☑

I refused to look at this list. I went back to the Crazy side and added

25 cats in a year ☑

I knew I was grasping. But I was resentful. The truth was, I needed someone to blame for Tibby's disappearance. I needed to be redirected from the uncomfortable realization that I was not enough for my cat and that he was keeping secrets from me. I'd thought he was pathologically shy, scared, unadventurous. Instead, he was taking up with strangers and spending time in bathhouses. By heavily weighting the Crazy column, I could shrug off this whole disappearance as an anomaly. "Of course it didn't make sense," I would tell my friends, twirling a finger next to my temple. "Crazy people don't make sense." And that would be that.

"We have a cat that comes and eats the food in the backyard, and sometimes he even comes into the house and sprays," Cat Stealer B said. "One day I just tied a note to his collar, asking his owners to please take better care of him, and to get in touch."

"You tied a note to his collar?" My face drained.

"Why, yes," she said.

And that's when my mental tally crumbled. My resentment, once hot, now boiled away. Denial slumped into a corner, drunk and passed out. My jealous side agreed to hide under a blanket. Suspicion was leaving, furtively boarding a Greyhound bus.

The little that was left of my good sense rallied.

The Cat Stealers weren't cat stealers. They were cat lovers. They went to great lengths for cats, and even tied notes to their collars. If they were crazy, it was normal, honest, hardworking cat crazy.

Ties notes to collars ☑

They were just like me.

HOMEGROWN CATNIP, A·K·A· KITTY CRACK

22.

We had found out what we had wanted to know: For five whole weeks, Tibby had camped nearby, eaten junk food, cavorted with strays, ignored my calls.

But why?

"It *is* confusing that Tibby was lost for so long, so close to home," Cat Whisperer A said.

"He wasn't lost," I said sadly. "It seems he simply wanted to leave for a while."

The Cat Whisperers nodded, their expressions sage. "If cats don't like where they're living, they'll just move into another house," they said, and explained that a cat named Blackie had arrived that way. His owners had snakes, in particular one large anaconda that roamed the house freely, and Blackie had had enough. The Cat Whisperers would deposit Blackie at his home, but within hours he would be back.

"Cats choose," Cat Whisperer B said, her voice gentle.

"But I don't have an anaconda," I whimpered.

Our meeting was drawing to a close when Wendy said, "Too bad we spent so much time on technology when we could have just talked to our neighbors." I hung my head, sheepish.

"Oh, I used to be really shy," Cat Whisperer A said, glancing at me with kind eyes. "And then ten years ago I became a schoolteacher and now I walk around the neighborhood and talk to people. But San Francisco can be a really cold place. It's a place where people stay indoors. We don't have any porches to hang out on, so it's hard to connect with people."

"Yes," I agreed, nodding at each Cat Whisperer in turn. "Well, I'm glad I finally talked to people. Or else I wouldn't have met both of you."

The Cat Whisperers stood to go. But one final thing puzzled me: Tibby had begun to eat at home almost immediately after the last round of flyers had been placed in mailboxes in the Suspicious Area. Could it be that the Cat Whisperers had seen our note and, shaken, withdrawn food? I began to form the question in my head.

But did it matter? It didn't change the fact that my cat had left of his own free will—and stayed away. I looked from one Cat Whisperer to the other. I frowned. If they'd actually received our flyer their whole story would be in doubt. If their story was in doubt . . .

I could hold on to some of my denial a little bit longer. Maybe Tibby had gone to Antarctica, or on rumspringa, or a kitty walkabout. Maybe he had been trapped, or kidnapped, or brainwashed.

Wendy looked at me. The Cat Whisperers looked at me. Wendy must have seen something in my face, because she narrowed her eyes.

What good was an obsessive quest if, when nearing the end, you forgot your obsession? No good at all. Wasn't this my last chance to hold on with scrabbling fingernails to the cat I thought (hoped) I once knew? Yes. Wasn't Truth the most important? Certainly. And so I dug deep, and after a moment, found Truth.

"Thank you for taking care of Tibby," I said. "Thank you so much."

23.

Operation Chasing Tibby was officially over. We gathered the GPS unit, the CatCam, the notes from my animal communications class, the flyers, and the laminated missive. We packed them into a box. We marked it "Tibia Stuff," and put it away on a shelf.

But there was one last thing that remained unexplained. Why had Tibby left?

In the hopes of coming closer to an answer, I decided on one final foray. I decided to visit the banya.

A truck was parked in front of the large three-unit house behind which, out of sight, lay the fabled ruins. I rang the doorbell. Then I waited. And waited. But no one answered. This cold, cold city, I thought. Just then a car pulled up, and a man got out. He lived there, yes. The banya? Sure, he would be happy to show me.

He opened the gate and we walked/crutched up an outside stairway that ran along the side of the house. At the top he pointed to what looked like a large run-down toolshed.

"There it is," he said.

The banya was indeed a kitty paradise. It had possum-size holes by which to enter and exit. The yard that surrounded it was a tangle of bushes. It was a world apart, a memory of bygone years, ignored by humans, perfect for animals.

OLD BANYA FLOOR PLAN

The banya was so strange—plopped down in an urban backyard like a fallen spaceship—that for a moment I let myself believe that maybe Tibby hadn't heard me calling him all those weeks. Perhaps there was an otherworldly force field around this area. Surely the dense shrubbery and brambles had acted as a sound buffer. But then a car alarm went off up the street, its monotonous shriek piercing and clear. I was yanked into the present. I thanked the neighbor and returned to my house.

I didn't need to turn on the computer and re-analyze the maps. I didn't need to scour the photos. I didn't need to have an animal-human conversation. Clear and bright as the pink of a kitty trail on a satellite map was this final truth: Tibby had just not wanted to be at home.

Every long-term relationship has its ups and downs. Last summer had definitely been a down. I had been a greasy-haired, foggy-eyed, catheter-wielding lump on the couch for months. My depression had leaked from every pore. My physical pain was palpable. Fibby, always possessive when I was home, had swatted and hissed at Tibby 24-7. Visitors had tramped in, bringing chocolate and sympathy. Family had stayed for weeks to help take care of me. Tibby had had enough.

Sometimes relationships end at this juncture. But sometimes they mend. I knew then that the point was not that Tibby had left.

The point was that Tibby had returned.

24.

Every quest is a journey, every journey is a story. Every story, in turn, has a moral. Here are seven possible morals of our story.

1. Technology is awesome. It's the wave of the future. Computers! GPS! Cat cams! Next time, buy the gizmo that orders pizza, irons shirts, and turns into a stun gun.

2. Don't rely on technology. It's fine for some things, but in our story, talking face-to-face with our neighbors proved the most beneficial and rewarding. For future quests, we recommend the most old-fashioned technology of all: the larynx-tongue-jawbone contraption.

3. I was once depressed, but then I got out in the world and I wasn't depressed anymore, just bonkers.

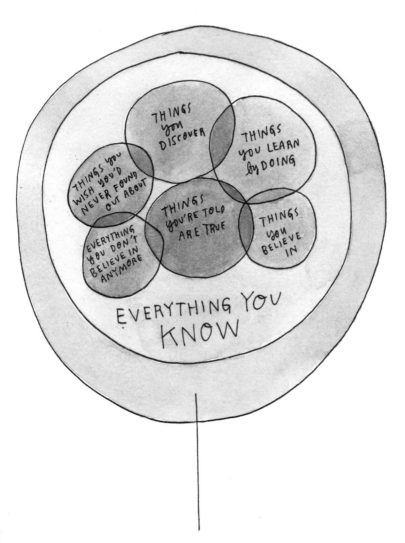

ALL THE THINGS
YOU WILL NEVER
UNDERSTAND

4. Bonkers is in the eye of the beholder.

5. Sooner or later everyone becomes a cat lover. Just ask Wendy.

6. You can never know your cat. In fact, you can never know anyone as completely as you want.

7. But that's okay, love is better.

Events in our household settled down. The crutches had finally been put away. Wendy moved in permanently. And Tibby showed no interest in wandering anymore. He still came and went freely, but he was never gone for long. Mostly he would saunter to a patio chair, look around, yawn, and stretch out underneath. He would close his eyes and nap for hours, dreaming dreams that I now accepted I would never know. Gradually he adjusted to being the only cat in the household, the sole recipient of laps and baby talk and cloying affection. He even seemed to like it.

Friskies were offered regularly.

Recently, Wendy proclaimed she wanted two kittens. We'll get them from the pound, she told me, and at the doddering old age of fourteen Tibby will have young feline companionship.

"Two kittens!" she exclaimed, her eyes shining. "Please?"

"Are you sure?" I asked.

I reminded her that she was about to embark on a journey of love, jealousy, and abject bewilderment.

She assured me that she could handle it. Yes, you can never know your kitties, she agreed, but the lesson she'd learned from Operation Chasing Tibby was that trust was paramount.

"I'll trust that they'll love me no matter what," she said, and I saw in her face all the hope and enthusiasm and future heartbreak of a new cat owner.

8. Trust is good, but there's always GPS.

GPS

Tibia
1994–2012
A Very Good Cat

A NOTE ON THE AUTHOR

Caroline Paul (carolinepaul.com) is the author of the novel *East Wind, Rain* and the memoir *Fighting Fire*. She lives with Wendy in San Francisco.

A NOTE ON THE ILLUSTRATOR

Wendy MacNaughton is an illustrator based in San Francisco. Her work has appeared in the *New York Times*, *Juxtapoz*, and *Print Magazine*. Her illustrated documentary series "Meanwhile" is published by the *Rumpus*. She lives with Caroline in San Francisco.